4

Len and Anne Frobisher

Heinemann
Halley Court, Jordan Hill, Oxford, OX2 8EJ
a division of Reed Educational and Professional Publishing Ltd
www.heinemann.co.uk

Heinemann is a registered trademark of Reed Educational and Professional Publishing Ltd

ISBN 0 435 20865 9 / 9780435208653 (Pupil Book)
ISBN 0 435 20873 X / 9780435208738 (Teacher's version)

09 08 07 06

Illustrated by Nick Schon

Cover illustration by Andrew Hunt

Typesetting and layout by DP Press Ltd, Sevenoaks, Kent

Printed by Ashford Colour Press Ltd, Gosport, Hants.

Contents

Number problems 1

place value; +10/100, –1/100

1 A book has the pages between 149 and 160 missing.

What is the tens digit of the missing pages? **5**

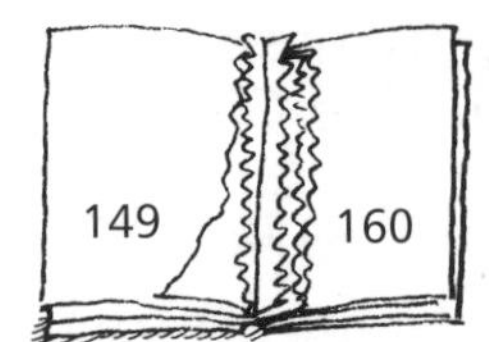

2 A shop has 297 toy monsters for sale.

Next day 10 more toy monsters are delivered.

How many does the shop have now? **307**

3 A post office has 836 first class stamps. It sells 100 of them.
How many first class stamps does it have left? **736**

4 There are 341 people in a Sports Centre.

Six groups each with 10 people come to the centre.

How many are in the centre now? **401**

5 Every day a bakery makes 850 tea cakes. One day it has a special order for an extra 100 tea cakes.

How many tea cakes does it make on that day? **950**

6 A lorry driver takes 900 bricks to a builder.
He leaves one of them on the lorry.

How many bricks does he give the builder? **899**

7

When you count on 3 tens from me and then count back 2 hundreds you get 400.

What am I?

570

Money problems 1

totals and change; convert £ to p; +1/100, −10/100

1 Gerry has saved £25.50 to buy the CD.

How much does he have left? **£15.50**

2 There are 549 pennies in the jar. Kate puts in 100 more.

How much in £s and p is in the jar? **£6.49**

3 Lauren has £13.99. She needs 1p more to buy a sweater.

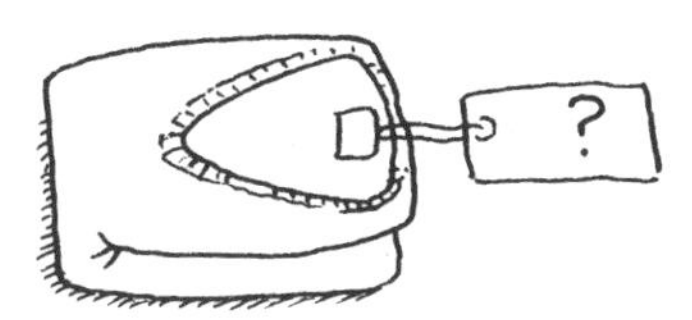

How much is the sweater? **£14.00**

4 Before the sale the TV was £307.

What is the price of the TV in the sale? **£297**

5 A washing machine is for sale at £463. Sam's mum offers the shop £100 less than the price.

How much does Sam's mum offer the shop? **£363**

6 A car is advertised for sale. Helen buys the car.

£949.50
£100
cashback

How much does the car cost Helen? **£849.50**

7 *If you start at me and count back 6 tens and then count on 4 hundreds you get to 740.*

What am I?

400

Number problems 2

one/two step problem
round to nearest 10/1

1 In a game Pam scores 40 points. In the next game she scores 70 points.

Pam's scores
game 1 40
game 2 70

How many points does she score altogether? **110**

2 Ten people share a prize of 3000 bottles of wine.

How many bottles does each get when it is shared equally? **300**

3 Each crate holds 8 bottles of milk. A class is sent 3 crates and 6 more bottles.

How many bottles does the class get? **30**

4 A show sells 753 tickets. Six people who got tickets do not come.

How many come to the show? **747**
Round the number to the nearest ten. **750**

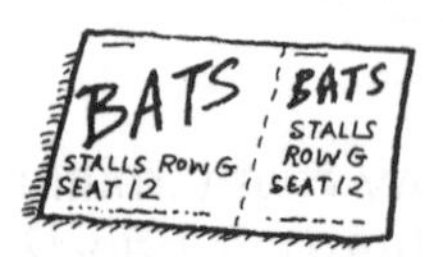

5 Each box holds 6 toy cars. Milly is given 5 boxes. She takes out all the cars and shares them equally between three garages.

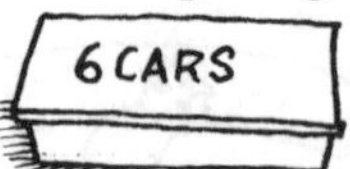

How many cars are in each garage? **10**

6 At the start of a match there are 5422 people in the ground. Six more come in. Before the end 9 leave.

How many are in the ground at the end of the match? **5419**

7

I am in both the 3 times and the 5 times-tables. The sum of my two digits is 3.

Wha
am I

3

Length problems 1

$\frac{1}{2}$, $\frac{1}{4}$, $\frac{3}{4}$, $\frac{1}{10}$ of 1 km in m and of 1 m in cm/mm; perimeter

1 Paula is in a 1 km cycle race.

How many metres has Paula to go when she has covered half the distance? **500 m**

2 A piece of wood is 1 m long. It is cut in to 10 equal lengths.

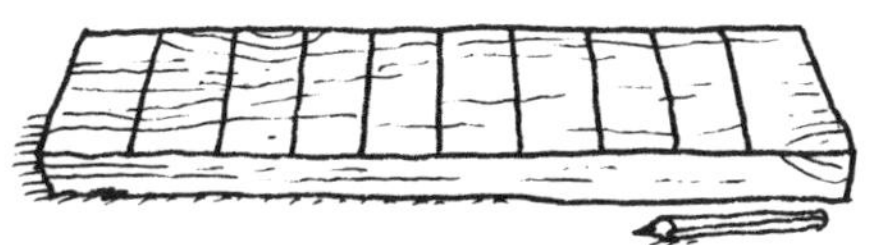

How many centimetres long is each piece? **10 cm**

3 Sara buys 1 metre of ribbon. She cuts one-quarter of it for a hair band.

How many millimetres long is her hair band? **250 mm**

4 A rectangular field is 1 km long and 500 m wide.

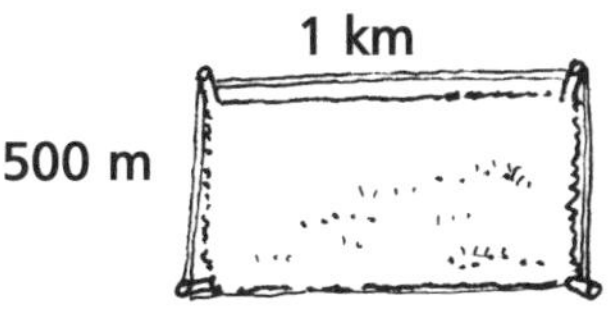

What is the perimeter of the field? **3 km**

5 A 1 km length of road has lamp posts every $\frac{1}{4}$ km.

How many metres is it from the start of the road to the $\frac{3}{4}$ km post? **750 m**

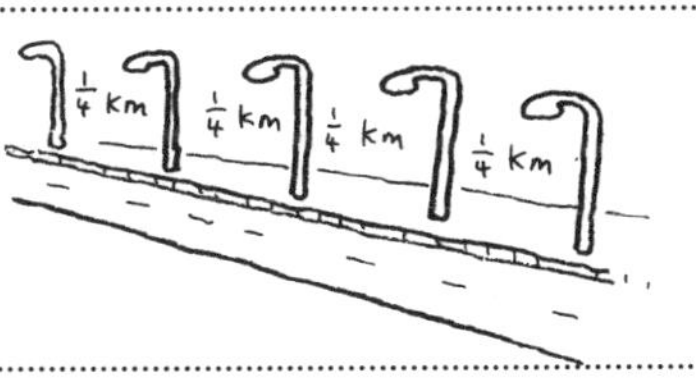

6 A room is 7.5 m long. Rod buys a carpet that is 8 m long.

How many centimetres longer than the room is the carpet? **50 cm**

7 *One-quarter of me is 10 metres.*

What am I?

40 m

Number problems 3

+/– facts to 20;
+/– 2D numbers

❶ At a party 17 stools are put around tables. Only 13 children come to the party.

How many stools are empty? 4

❷ There are 12 people in a bus queue. Another 5 join them before 8 people get on a bus.

How many people are still in the queue? 9

❸ There are 11 people in a burger bar. Eight more come in.

How many are in the burger bar now? 19

❹ Marc has 53 adventure cards. He buys another 25 cards.

How many cards does he have now? 78

❺ In a plantation there are 98 trees. Loggers cut down 36 of them and plant 23 more.

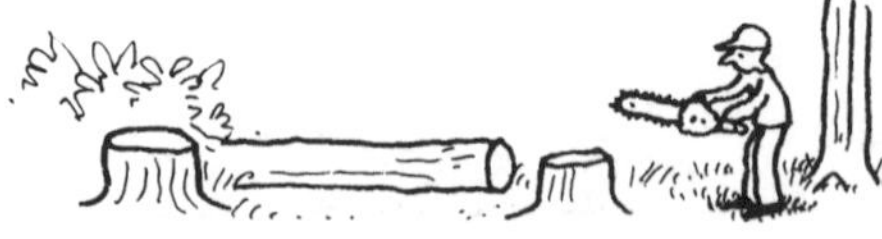

How many trees are left in the plantation? 85

❻ There should be 87 passengers on a plane. A count shows that there are only 75.

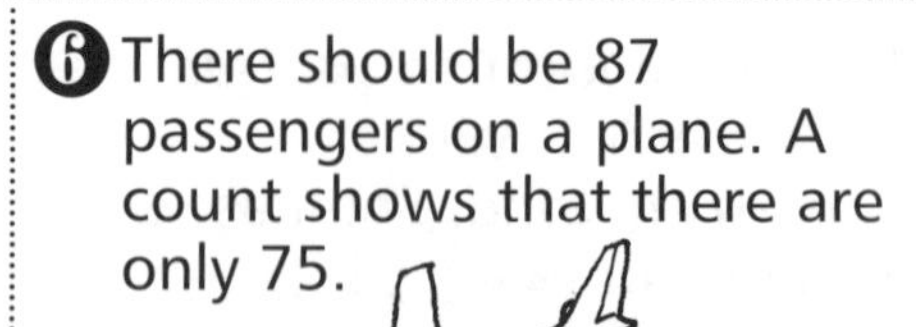

How many passengers are missing? 12

❼

I have two digits. I am between 16 and 24. I am one-quarter of a multiple of 10.

What am I?

20

Measures problems 1

×/÷ of measures;
×/÷ 2 and 3

1 Each bag of apples weighs 3 kg 200 g.
A bag of oranges weighs double this.

How much does the bag of oranges weigh? **6 kg 400 g**

2 A bottle holds 2 L 500 mL.

How much do two bottles hold? **5 L**

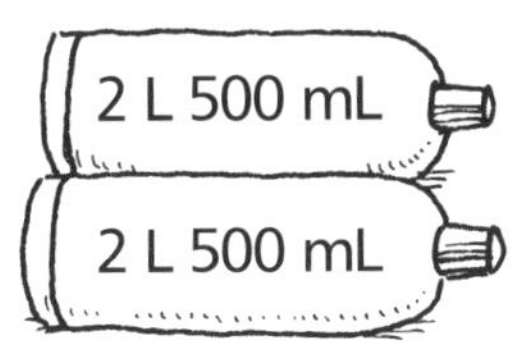

3 Emma folds a 60 cm length of wool into three equal pieces.

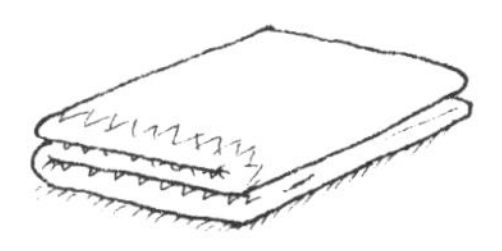

How long is each piece of wool? **20 cm**

4 A slow train from Walton to Hesford takes 1 hr 20 min. A fast train takes half the time.

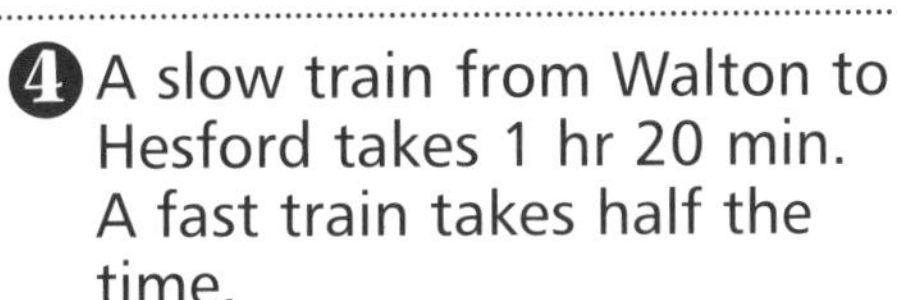

How long does the fast train take? **40 min**

5 Alex has a knitting pattern for a scarf. It will be 1 m 20 cm long. She has already done half of it.

How much has she still to do? **60 cm**

6 Ross runs 100 metres in 15 seconds. His smaller sister takes three times as long.

How long does his sister take? **45 sec**

7

I am in the 2 times-table and the 3 times-table. The difference between my two digits is 5.

What am I?

72

Review problems 1

1. On which day is Halford Show open longer? **Saturday**
   For how many hours longer? **3 hours**
2. On Saturday, 100 adults and 50 children visit the show. On Sunday 200 adults and 100 children come.
   How many people visit the show altogether? **450**
   What is the total amount collected for entry to the show for the two days? **£885**
3. The fee to show items is £1.50. There are 400 exhibitors.
   How much is paid in fees? **£600**
4. A raffle is held. Tickets are 5 for £1. Out of 600 tickets, 500 are sold.
   How many tickets are not sold? **100**
   How much money is collected in the raffle? **£100**
5. Apples are shown in groups of 4. Eighteen people display a group of apples.
   How many apples are shown altogether? **72**
6. Altogether 96 tomatoes are on display. Each exhibitor shows 3 tomatoes.
   How many exhibitors show a group of tomatoes? **32**
7. There are 10 cucumbers on display. Each measures 26 cm.
   How long in m and cm would the ten cucumbers be if put end to end? **2 m 60 cm**
8. The heaviest pumpkin weighs 14 kg. The lightest weighs 11 kg.
   What is the difference in weight between the heaviest and lightest pumpkins? **3 kg**
9. There are 20 different classes of fruit and vegetables. In each class a 1st, 2nd and 3rd prize is given.
   How many prizes are awarded altogether? **60**

Number problems 4

odd/even; number puzzles; +/– facts to 20

1 Jenny buys 6 plants for the house and 9 plants for the garden.

How many plants did she buy altogether? **15**

2 Amy thinks of two numbers. The sum of the two numbers is 10. The product of the two numbers is 21.

What are Amy's two numbers? **7 and 3**

3 The houses on one side of a street have odd numbers. Ario lives at number 127. Dale lives 10 houses up from Ario.

What is Dale's number? **147**

4 There are 15 cakes on a plate. Four children eat 8 of the cakes.

How many cakes are left on the plate? **7**

5 In a race there are 18 hot air balloons. Two of the balloons do not take off. Two balloons have to land.

How many balloons are still in the air? **14**

6 The 15 doors down one side of a corridor in a hotel are even numbers. The first door is 102.

What number is the last door? **130**

7

I am an even number. The sum of my two digits is 6. Half of me is an odd number.

Wha
am I

4

Money problems 2

totals and change; count on/back in 10s/100s

1 Mary empties her money box. She counts out 75p. She still has seven more 10ps to count.

How much money did she have in her money box? **£1.45**

2 Jake has £3.82 in his pocket. The rest of his money he collects into 5 groups of 100p. He counts all his money.

How much has Jake? **£8.82**

3 Tino buys a comic for £1.30. He pays with a £2 coin. The shopkeeper counts out the change in 10p coins.

How many 10p coins does the shopkeeper give Tino? **7**

4 Pam's mum has her garden paved. It costs £887. She gives the gardener £87 and then some £100 notes.

How many £100 notes does she give the gardener? **8**

5 A model car kit costs £7.30. Holly gives the shopkeeper £5 and counts out the rest in 10p coins.

How many 10p coins does she count out? **23**

6 Joe buys a pair of socks for £2.70. Joe counts out 10p coins until he has £2 left to pay.

How many 10ps does he count out? **7**

7

If you start at me and count on 4 tens and then count back 4 hundreds you get to 0.

What am I?

360

Number problems 5

doubles/halves; TU×U, TU÷U;

1 After 5 games Gary has scored 9 points. In the next 5 games he doubles his number of points and then doubles that in the next 2 games.

What is Gary's total after 12 games? **36**

2 Mae shares 39 strawberries equally into 3 bowls.

How many strawberries are in each bowl? **13**

3 A block of flats has 6 floors. The builders put the same number of windows in each floor. Altogether they put in 72 windows.

How many windows are in each floor? **12**

4 At a concert there are 200 people. Half of them are children. Half of the children are boys.

How many boys are at the concert? **50**

5 Between them, 5 children sell 125 tickets for the Autumn Fair. Each child sells the same number of tickets.

How many tickets does each child sell? **25**

6 There are 6 classes in a school. Each class has 28 children.

How many children are there in the school? **168**

7

I am a multiple of 5, but not a multiple of 10. My units and hundreds digits are the same. The sum of my three digits is 11.

What am I?

515

Number problems 6

mixed numbers; fractions of quantities; +/– 2D numbers

❶ A shop has 8 pizzas. Each pizza is cut into quarters. The shop sells 25 quarter pieces.

How many quarter pizzas are left? **7**

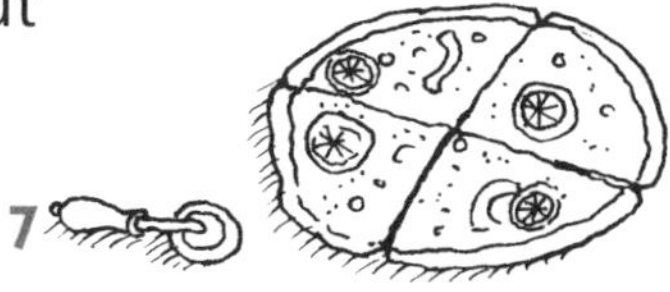

❷ A potter makes 72 fancy teapots. She sells 55 of them.

How many teapots has she left for sale? **17**

❸ A home for cats has 47 strays. Then 18 more are brought in.

How many cats are in the home? **65**

❹ A zoo has 27 new animals. One-third of them are monkeys.

How many monkeys arrive at the zoo? **9**

❺ Bibi has read 29 pages of her new book. When she goes to bed she reads another 56 pages.

How many pages has she read altogether? **85**

❻ A farmer has 86 geese. She sells 39 of them at market.

How many geese has she left? **47**

❼

When you add 29 to me and then subtract 13 from the answer you get 48.

What am I?

32

Time problems 1

time to 1 minute; a.m. and p.m.; doubles/halves

1 A clock shows 8:44. The clock is 15 minutes fast.

What is the correct time? **8:29**

2 **It takes Javed 1 min 10 sec** to count to 100. When he counts from 100 to 200 it takes him twice as long.

How long does it take Javed to count from 100 to 200? **2 min 20 sec**

3 A bus takes 1 hr 50 min to get from Donby to North and 50 min to get from North to Peters.

How long does it take the bus to get from Donby to Peters? **2 hr 40 min**

4 There are four swimmers in a relay race team. Each swimmer takes 35 seconds.

How long does it take for the team to swim the relay? **2 min 20 sec**

5 A bus journey takes 6 hr 30 min. A train does the journey in half the time.

How long does it take the train? **3 hr 15 min**

6 Each morning Year 4 are in school from 8:50 a.m. to 12:10 p.m. They spend a quarter of this time doing Maths.

For how long do they do Maths? **50 min**

7

If you double me and halve the answer you get 81.

What am I?

81

Number problems 7

small differences;
HTU+/–TU;
2, 3 and 4× and ÷ facts

1 In a women's cricket match Australia score 503 runs and South Africa score 499 runs.

How many more runs do Australia score than South Africa? **4**

Australia	503
South Africa	499

2 In a rugby match Hampshire score 17 points, but Somerset score double that.

How many points do Somerset score? **34**

3 At an afternoon pantomime there are 389 people. In the evening there are 76 more than in the afternoon.

How many people were at the evening show? **465**

4 Rod has 18 animal cards. He shares them equally between himself and his two friends.

How many cards does each friend get? **6**

5 **A shop sells packets of Samson chocolate biscuits.**

How many biscuits are in 4 packets? **36**

6 In West Primary School there are 325 children. Of them, 68 go on a school trip.

How many children do not go on the trip? **257**

7

I am two odd numbers. Neither of my numbers is 1. The product of my numbers is 35.

What am I?

7, 5

Review problems 2

1. Mrs Halt wants 60 baubles.
   How many boxes of baubles does she need to buy? **6**
   How much will they cost her? **£15**

2. Mr Bryan pays £6.50 for some single baubles and figures.
   How many does he buy? **5**

3. Walter is going to buy 100 cards.
   How much would it cost him if he bought them in boxes of 20? **£18**
   How much would it cost him if he bought them in boxes of 50? **£15.98**
   Which way of buying them is cheaper? **50s**
   How much cheaper? **£2.02**

4. What is the cost of each cracker in a box of 10? **52p**
   What is the cost of each cracker in a box of 20? **50p**

5. Sally buys 20 lengths of tinsel; the 20 lengths fit exactly round a square window.
   What is the perimeter of the window? **20 m**
   What is the length of a side of the window? **5 m**

6. Mary buys 1 box of baubles and 5 lengths of tinsel. She pays £10.
   What is the cost of 1 metre of tinsel? **£1.50**

7. Mr Jones wishes to give each child in Years 3 and 4 a cracker. There are 70 children altogether.
   Which boxes could he buy? **3 x 20 + 1 x 10, or 7 x 10**
   How much would it cost him? **£35.20 or £36.40**

8. How many centimetres taller is the 2 m tree than the 1.5 metre tree? **50 cm**

Number problems 8

more;
×/÷ 10; +/– facts to 20;
–ve numbers

1 There were 7822 people watching Bradley United and 7550 watching Manford City play football.

How many more watched Bradley than Manford? **272**

2 A library has 780 children's books. The books are divided equally between 10 shelves.

How many books are on each shelf? **78**

3 One morning a school has 18 visitors. Only nine of them stay for the afternoon.

How many visitors leave at lunchtime? **9**

4 Pat has 14 ice creams in her freezer. At the supermarket she buys another packet of 6 ice creams for the freezer.

How many ice creams will be in the freezer? **20**

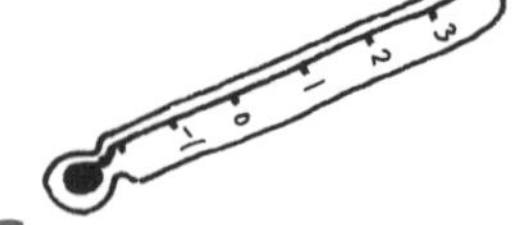

5 The temperature at midnight was 3°C. By 1 a.m. the temperature had dropped by 4°C.

What was the temperature at 1 a.m? **–1°C**

6 There are 10 biscuits in a packet.

How many biscuits are in the large box? **1500**

7

I am between 3591 and 5807.
I am an even multiple of 1000.

What am I?

4000

Money problems 3

totals and change;
+/– facts to 20;
3, 4 and 5× and ÷ facts

1 Sean buys 3 computer games.

How much do they cost him? **£24**

2 Kylie buys some toy farm animals for £8 and a toy tractor for £7.

What change does she get from a £20 note? **£5**

3 Five children go on a train. The total cost is £45.

What is the cost for each child? **£9**

4 Jane buys the pair of shoes and the pair of sandals.

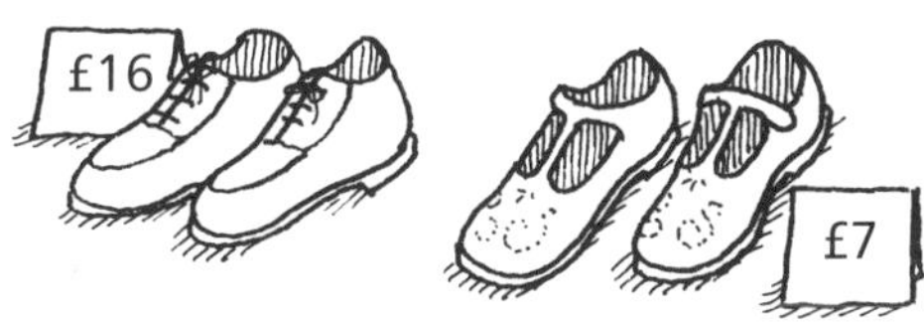

How much more do the shoes cost than the sandals? **£9**

5 For his 9th birthday Perry gets £10 from each of his four grandparents.

How much does Perry get altogether? **£40**

6 Mr and Mrs Rose take their four children to the cinema.

How much does it cost them? **£27**

7

I am less than 100.
I am a multiple of 3, 4 and 5.

What am I?

60

Number problems 9

+ several numbers; HTU+/–TU, HTU+/–HTU, –2D numbers

1 Deja threw a pair of dice four times. He scored 2, 3, 5 and 8.

What was his total score in the four throws? **18**

2 At a play there were 456 people. In the theatre 59 seats were empty.

How many seats did the theatre have? **515**

3 There were 662 people waiting at an airport for planes. Of these, 318 got on an Airbus.

How many were still waiting? **344**

4 Adam has 80 stickers in a book. He finds another 30 stickers in a drawer and then buys another 70.

How many has he altogether? **180**

5 Gemma plants 64 bulbs. Only 47 of them flower.

How many bulbs do not flower? **17**

6 At a school there are 185 children in the infants and 242 in the juniors. Because of illness 77 children do not come to school.

How many children are in school? **350**

7

I am more than 20. I am the sum of three different odd single-digit numbers.

What am I?

21

Time problems 2

sec/min/hr;
+/–2D numbers;
×2 and 5;
÷ by 10

1 When training an athlete runs two laps of a track at the same speed. He runs each lap in 50 sec.

How long does it take him to do the two laps?
1 min 40 sec

2 A train takes 37 minutes to reach the first station. It takes another 48 minutes to get to the second station.

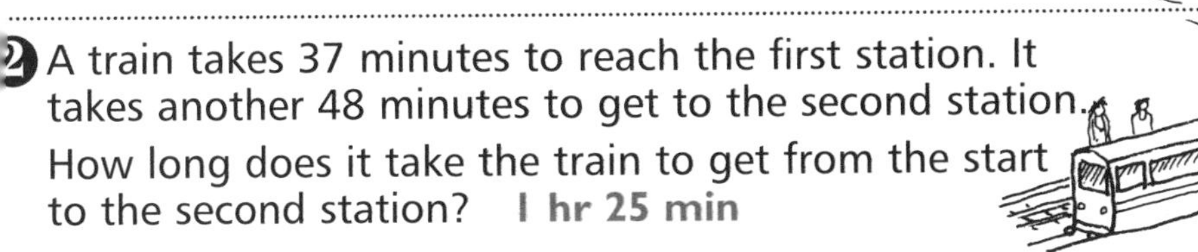

How long does it take the train to get from the start to the second station? **1 hr 25 min**

3 Tom does 10 step-ups in 1 minute. Each step-up takes the same time.

How many seconds does each step-up take? **6 sec**

4 Laura runs and walks for 1 hr 34 min. The walking takes her 56 minutes.

For how many minutes does she run? **38 min**

5 A snail takes 20 minutes to cover 1 metre.

How long will it take the snail to cover 5 metres at the same speed? **1 hr 40 min**

6 John went to bed at 7:30 p.m. and woke up at 6:15 a.m. the next morning.

For how long was he asleep?
10 hr 45 min

7

I am two numbers. The sum of my numbers is 100. The difference between my two numbers is 30.

What am I?

35, 65

Weight problems 1

$\frac{1}{2}$, $\frac{3}{4}$, $\frac{1}{10}$ of 1 kg in g; decimals; halves; ÷2 and 3

1 James bakes some buns. He uses $\frac{1}{10}$ kg of sugar.

How many grams of sugar does James use? **100 g**

2 A van is carrying 800 jars of jam. Each jar weighs 1 kg. The driver delivers half of the jars.

What is the weight of the jars that are left? **400 kg**

3 Jill's 9th birthday cake weighs 1 kg 800 g. It is cut into two equal pieces.

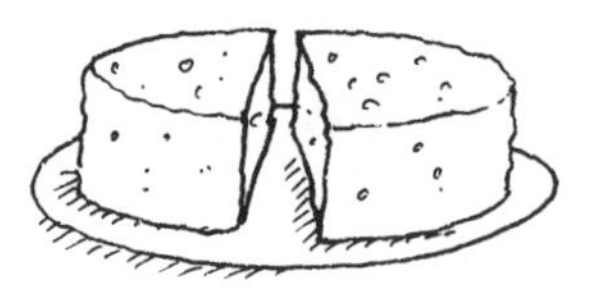

What is the weight of one piece? **900 g**

4 A chocolate roll weighs 1 kg. It is cut into four equal pieces. Reg eats one piece.

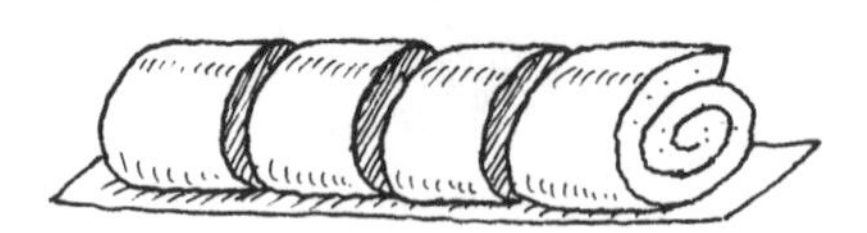

How many grams of the cake are left? **750 g**

5 Milly weighs a $\frac{1}{2}$ kg ball of dough and then another ball which weighs $\frac{1}{4}$ kg.

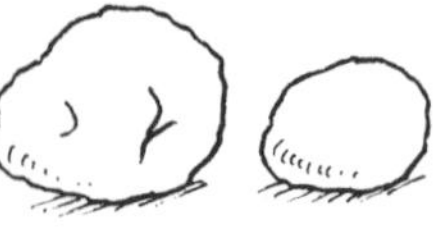

She rolls the two pieces together.

What is the weight of the dough as a decimal of a kg? **0.75 kg**

6 Francis makes 900 g of pizza dough. He divides it up into three equal pieces.

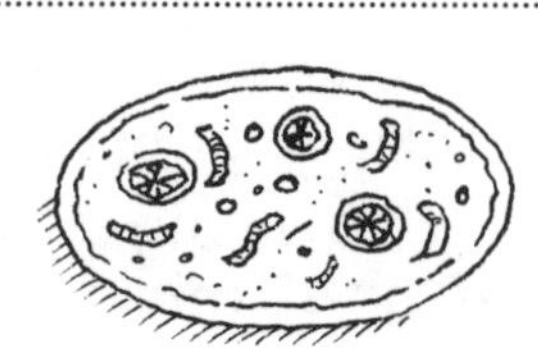

How heavy is each piece? **300 g**

7

Three-quarters of me is 300 g.

What am I?

400 g

Teachers' Notes

Contents

Introduction

The Pupil's books

Word Problems, a series of four books, one for each of the years 3 to 6, helps develop children's ability to solve number problems in a variety of contexts. Each page of word problems may be used to support numeracy lessons taught earlier in a week or for homework. The books provide weekly practice of word problems, as described in the National Numeracy Strategy (NNS) *Framework* and are designed to match the weekly structure within the NNS "sample medium-term plans". The style of questions reflects the examples that appear in the National Curriculum tests.

The books contain two types of page:

Topic pages: Each page has six word problems which are devoted solely to one Topic (Number, Money, Time, Length, Weight, Capacity and Measures). The overall mathematical content of all the Topic pages is listed on page iii. The mathematical content of the word problems on each page is listed in the lozenge at the top of the page.

Question 7 on each Topic page is a number puzzle, which gives children practice in Reasoning with Number using a variety of number properties. These questions are similar to those in the *Framework* and the National Curriculum tests.

Review pages look back at the mathematical content of previous Topic pages. Some Review pages are double page spreads, with the left page having a scene with information that is required to answer questions, and the right page asking questions about the scene.

Illustrations are used to tune children in to the 'real life' context of the word problem. Each is part of the problem and it is important that children look at the illustrations closely, they may contain information that is also in the word problem itself. On some occasions an illustration will contain information that is not in the text of the word problem, but is essential in order to solve the problem.

Answers to word problems are printed in red at the side of each question in the pupils' pages of the Teachers' notes.

Helping children solve word problems

Here are some suggestions for helping children develop a strategy for solving word problems.

- Make sure children read a question carefully and not merely search for key words such as 'altogether' which they think, sometimes incorrectly, tell them what to do with the numbers.
- Encourage children to close their eyes and picture the context of a problem and any actions that are performed with/on 'objects' in the context.
- Allow children to talk with other children about a problem and to ask themselves:
  - 'What do I have to find out?'
  - 'What do I know that will help me find out?'
  - 'What do I have to do with what I know to find out?'

It is also important as part of a word problem solving strategy that children develop:

- the skill of recognising information that is helpful and that which is not helpful when trying to solve a word problem.
- the ability to choose and use appropriate operations to solve word problems.

These can only develop with experience of solving many word problems over a long period of time.

Teaching plans

To enable you to integrate ***Word Problems*** into your medium-term teaching plans pages iv to vi show the relationship between the Topics in ***Word Problems*** and the Sample Medium-term plans suggested by the NNS. Pages vii and viii show how the Topics in ***Word Problems*** relate to Mathematics in the National Curriculum in Wales Programme of Study and the similar Programme of Study for Northern Ireland.

Summary of mathematical content

page	topic	Mathematical content
4	Number 1	place value; +10/100; −1/100
5	Money 1	totals and change; convert £ to p; +1/100; −10/100
6	Number 2	one/two step problems; round to nearest 10/100
7	Length 1	1/2; 1/4; 3/4; 1/10 of 1 km in m and of 1m in cm/mm; perimeter
8	Number 3	+/− facts to 20; +/− 2D numbers
9	Measures 1	÷/4 of measures; ×/÷ 2 and 3
10–11	**Review 1**	**review of previous content**
12	Number 4	odd/even; number puzzles; +/− facts to 20
13	Money 2	totals and change; count on/ back in 10s/100s
14	Number 5	doubles/halves; TU×U; TU÷U
15	Number 6	mixed numbers; fractions of quantities; +/− 2D numbers
16	Time 1	time to 1 minute; a.m. and p.m.; doubles/halves
17	Number 7	small differences; HTU+/−TU; 2; 3 and 4 × and ÷ facts
18–19	**Review 2**	**review of previous content**
20	Number 8	more; ×/÷ 10; +/− facts to 20; −ve numbers
21	Money 3	totals and change; +/− facts to 20; 3, 4 and 5 × and ÷ facts
22	Number 9	+ several numbers; HTU+/−TU; HTU+/−HTU; − 2D numbers
23	Time 2	sec/min/hr; +/− 2D numbers; × 2 and 5; ÷ by 10
24	Weight 1	1/2; 3/4; 1/10 of 1kg in g; decimals; halves; ÷ 2 and 3
25	Measures 2	area; ×8 and 10; ÷4 and 10
26–27	**Review 3**	**review of previous content**
28	Number 10	−ve numbers; doubles/halves; + 2D numbers; general statements
29	Money 4	÷ £s by 2, 4, 5 or 10; − 2D numbers; multiples of 50 total 1000
30	Number 11	× 9 and 11; TU×U; +/− facts to 20
31	Number 12	fractions total 1; equivalent fractions; 2, 4, 10×; ÷ 5
32	Number 13	multi–step
33	**Review 4**	**review of previous content**
34	Number 14	× 100; +/− 10/100/1000
35	Money 5	+/− 10/100/100; doubles
36	Number 15	facts to +/−; +/− nearest multiple of 10; + multiples of 10; halves
37	Capacity 1	1/2; 1/4; 1/10 of 1L in mL; pints; +/− facts to 20
38	Measures 3	pairs total 100; multiples of 50 total 1000
39	Measures 4	angles; ×/÷ 10/100
40–41	**Review 5**	**review of previous content**
42	Number 16	+/− facts to 20; × 2, 3, 4 and 5
43	Money 6	× 2 and 4; ÷ 5 and 10; round up/down after ÷
44	Number 17	× 9 and 11; + 2D numbers; ÷ 2 and 4; TU÷U; round up/down after ÷
45	Number 18	proportion; − 2D numbers; × 6 and 8; 1/2 as decimal
46	Time 3	timetables; calendars; × 3, 4 and 9
47	Number 19	+/− 2D numbers; × 3 and 5; HTU+/−HTU
48	**Review 6**	**review of previous content**

Word Problems and the National Numeracy Strategy
Sample medium–term plans

AUTUMN			
Sample medium-term plans		*Word Problems*	
Unit	**Topic**	**Pages**	**Topic**
1	Place value; ordering; rounding Reading numbers from scales	4	Number 1
2–3	Understanding + and – Mental calculation strategies (+ and –) Paper and pencil procedures (+ and –) Money and 'real life' problems Making decisions; checking results	5–6	Money 1 Number 2
4–6	Measures; including problems Shape and space Reasoning about shapes	7–9	Length 1 Number 3 Measures 1
7	**Assess and review**	**10–11**	**Review 1**
8	Properties of numbers Reasoning about numbers	12	Number 4
9–10	Understanding × and ÷ Mental calculation strategies (× and ÷) Pencil and paper procedures (× and ÷) Money and 'real life' problems Making decisions; checking results	13–14	Money 2 Number 5
11	Fractions and decimals	15	Number 6
12	Understanding + and – Mental calculation strategies (+ and –) Paper and pencil procedures (+ and –) Time, including problems	16	Time 1
13	Handling data	17	Number 7
14	**Assess and review**	**18–19**	**Review 2**

SPRING			
Sample medium-term plans		*Word Problems*	
Unit	**Topic**	**Pages**	**Topic**
1	Place value; ordering; rounding Reading numbers from scales	20	Number 8
2–3	Understanding + and – Mental calculation strategies (+ and –) Pencil and paper procedures (+ and –) Money and 'real life' problems Making decisions; checking results	21 22	Money 3 Number 9
4–6	Measures, and time, including problems Shape and space Reasoning about shapes	23 24 25	Time 2 Weight 1 Measures 2
7	**Assess and review**	**26–27**	**Review 3**
8	Properties of numbers Reasoning about numbers	28	Number 10
9–10	Understanding × and ÷ Mental calculation strategies (× and ÷) Pencil and paper procedures (× and ÷) Money and 'real life' problems Making decisions, checking results	29 30	Money 4 Number 11
11	Fractions and decimals	31	Number 12
12	Handling data	32	Number 13
13	**Assess and review**	**33**	**Review 4**

SUMMER			
Sample medium–term plans		*Word Problems*	
Unit	**Topic**	**Pages**	**Topic**
1	Place value, ordering, rounding Reading numbers from scales	34	Number 14
2–3	Understanding + and – Mental calculation strategies (+ and –) Pencil and paper procedures (+ and –) Money and 'real life' problems Making decisions; checking results	35–36	Money 5 Number 15
4–6	Measures, including problems Shape and space Reasoning about shapes	37–39	Capacity 1 Measures 3 Measures 4
7	**Assess and review**	**40–41**	**Review 5**
8	Properties of numbers Reasoning about numbers	42	Number 16
9–10	Understanding × and ÷ Mental calculation strategies (× and ÷) Pencil and paper procedures (× and ÷) Money and 'real life' problems Making decisions, checking results	43–44	Money 6 Number 17
11	Fractions and decimals	45	Number 18
12	Understanding + and – Mental calculation strategies (+ and –) Pencil and paper procedures (+ and –) Time, including problems	 46	Time 3
13	Handling data	47	Number 19
14	**Assess and review**	**48**	**Review 6**

Word Problems and the National Curriculum in Wales

Using and Applying Mathematics
U1. Making and Monitoring Decisions to Solve Problems
U2. Developing Mathematical Language and Communication
U3. Developing Mathematical Reasoning
Number
N1. Understanding Number and Place Value
N2. Understanding Number Relationships and Methods of Calculation
N3. Solving Numerical Problems
Shape, Space and Measures
S2. Understanding and Using Properties of Position and Movement
S3. Understanding and Using Measures

Word Problems		*Relevant Sections of the National Curriculum Programme of Study*							
page	**topic**	**U1**	**U2**	**U3**	**N1**	**N2**	**N3**	**S2**	**S3**
4	Number 1	x	x	x	x	x	x		
5	Money 1	x	x	x			x		
6	Number 2	x	x	x	x	x	x		
7	Length 1	x	x	x	x		x		x
8	Number 3	x	x	x		x	x		
9	Measures 1	x	x	x			x		x
10–11	**Review 1**	x	x	x		x	x		x
12	Number 4	x	x	x	x	x	x		
13	Money 2	x	x	x	x		x		
14	Number 5	x	x	x		x	x		
15	Number 6	x	x	x	x	x	x		
16	Time 1	x	x	x			x		x
17	Number 7	x	x	x		x	x		
18–19	**Review 2**	x	x	x		x	x		x
20	Number 8	x	x	x	x	x	x		
21	Money 3	x	x	x			x		
22	Number 9	x	x	x		x	x		
23	Time 2	x	x	x	x		x		x
24	Weight 1	x	x	x	x		x		x
25	Measures 2	x	x	x	x		x		x
26–27	**Review 3**	x	x	x		x	x		x
28	Number 10	x	x	x	x	x	x		
29	Money 4	x	x	x	x		x		
30	Number 11	x	x	x		x	x		
31	Number 12	x	x	x	x	x	x		
32	Number 13	x	x	x		x	x		
33	**Review 4**	x	x	x		x	x		x
34	Number 14	x	x	x	x	x	x		
35	Money 5	x	x	x			x		
36	Number 15	x	x	x		x	x		
37	Capacity 1	x	x	x	x		x		x
38	Measures 3	x	x	x			x		x
39	Measures 4	x	x	x			x	x	x
40–41	**Review 5**	x	x	x		x	x		x
42	Number 16	x	x	x		x	x		
43	Money 6	x	x	x	x		x		
44	Number 17	x	x	x	x	x	x		
45	Number 18	x	x	x	x	x	x		
46	Time 3	x	x	x			x		x
47	Number 19	x	x	x		x	x		
48	**Review 6**	x	x	x		x	x		x

Word Problems and the National Curriculum in Northern Ireland

PROCESSES IN MATHEMATICS
P1. Using Mathematics
P2. Communicating Mathematically
P3. Mathematical Reasoning
SHAPE AND SPACE
S2. Position, Movement and Direction

NUMBER
N1. Understanding Number and Number Notation
N2. Patterns, Relationships, and Sequences
N3. Operations and their Applications
N4. Money
MEASURES (M)

Word Problems		*Relevant Sections of the National Curriculum Programme of Study*								
page	**topic**	**P1**	**P2**	**P3**	**N1**	**N2**	**N3**	**N4**	**M**	**S2**
4	Number 1	x	x	x	x	x				
5	Money 1	x	x	x				x		
6	Number 2	x	x	x	x		x			
7	Length 1	x	x	x	x				x	
8	Number 3	x	x	x			x			
9	Measures 1	x	x	x					x	
10–11	**Review 1**	x	x	x			x	x	x	
12	Number 4	x	x	x	x	x	x			
13	Money 2	x	x	x	x			x		
14	Number 5	x	x	x		x	x			
15	Number 6	x	x	x	x	x	x	x		
16	Time 1	x	x	x		x			x	
17	Number 7	x	x	x			x			
18–19	**Review 2**	x	x	x			x	x	x	
20	Number 8	x	x	x	x	x	x			
21	Money 3	x	x	x				x		
22	Number 9	x	x	x		x	x			
23	Time 2	x	x	x	x				x	
24	Weight 1	x	x	x	x				x	
25	Measures 2	x	x	x	x				x	
26–27	**Review 3**	x	x	x			x	x	x	
28	Number 10	x	x	x		x	x			
29	Money 4	x	x	x	x			x		
30	Number 11	x	x	x			x			
31	Number 12	x	x	x	x		x			
32	Number 13	x	x	x			x			
33	**Review 4**	x	x	x			x	x	x	
34	Number 14	x	x	x	x	x	x			
35	Money 5	x	x	x		x				
36	Number 15	x	x	x		x	x	x		
37	Capacity 1	x	x	x	x				x	
38	Measures 3	x	x	x					x	
39	Measures 4	x	x	x					x	x
40–41	**Review 5**	x	x	x			x	x	x	
42	Number 16	x	x	x			x			
43	Money 6	x	x	x	x			x		
44	Number 17	x	x	x	x	x	x			
45	Number 18	x	x	x	x		x			
46	Time 3	x	x	x					x	
47	Number 19	x	x	x			x			
48	**Review 6**	x	x	x			x	x	x	

Measures problems 2

×8 and 10;
÷4 and 10; area

1 A bottle of lemonade holds 2 litres.

How many litres do 8 bottles hold? **16 L**

2 The distance around a model railway track is 4 m 50 cm. An engine goes round the track 10 times.

What distance does the engine travel? **45 m**

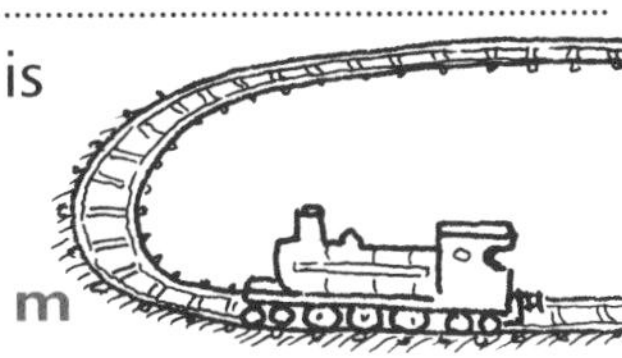

3 A 1 m 20 cm length of timber is cut into four equal lengths.

How many centimetres long is each piece? **30 cm**

4 A plane takes 1 hr 10 min to fly from London to Newcastle. A train takes 3 hr 20 min to do the same journey.

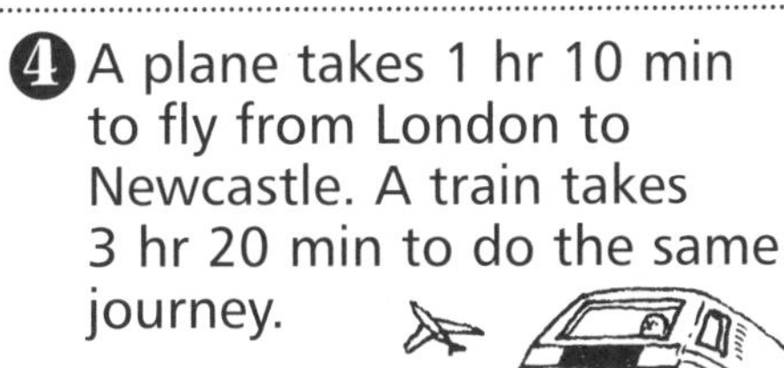

How much longer than the plane does it take the train? **2 hr 10 min**

5 Ten children each make a ball of the same weight. The total weight of the ten balls is 2 kg.

How many grams is the weight of one ball? **200 g**

6 A rectangular garden is paved with 1 m square paving stones. The garden is 15 m long and 10 m wide.

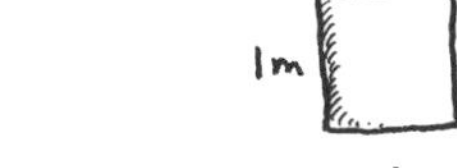

How many paving stones are used? **150**

7

My four digits are the same. The sum of my four digits is 8.

What am I?

2222

Review problems 3

Richman Primary School
Hardcastle Lane
Wensford
WD13 6GH

20th June 2002

Dear Parent/Guardian,

On Monday, 3rd July Year 4 are going on an educational visit to Apple Tree Farm. Adult helpers are encouraged to join the children.

The cost is £3.50 for each child and £5.00 per adult.

Buses will leave the school at 9:05 and we plan to return at 3:10.

Each child will need a packed lunch and up to £1 spending money only.

Please complete the attached slip and return it to school before Monday, 27th June.

Yours sincerely,

Ben Watson
Headteacher

review of previous content

1. How many days are there from the date of the letter to the date of the school trip? **13 days**

2. **There are 30 children in Year 4.**
   What is the total cost for the 30 children? **£105**

3. **Altogether 30 children and 15 adult helpers go on the trip.**
   How much money is paid to the school? **£180**

4. **As well as the 15 adult helpers, 4 teachers go on the trip.**
   How many adults altogether go on the trip? **19**

5. **The bus has 55 seats.**
   How many seats are empty? **6**

6. **The bus leaves 15 minutes late.**
   At what time does the bus leave school? **9:20**

7. **The bus leaves Apple Tree Farm at 2:30. It arrives back at school** 5 minutes early.
   How long does the bus take to get back? **35 min**

8. **The distance from the school to the farm is 19 miles.**
   How far is it there and back? **38 miles**

9. **One-half of the 30 children bring 50p spending money.**
   How many children bring 50p? **15**

10. **One-half of the 30 children bring 50p, and one-fifth bring 70p.**
    The rest bring £1.
    How many children bring £1? **9**

Number problems 10

–ve numbers; doubles/halves; +2D numbers; general statements

1 At 8 p.m. the temperature is 4°C. Every hour the temperature drops by 2°C until midnight.

What is the temperature at midnight? **–4°C**

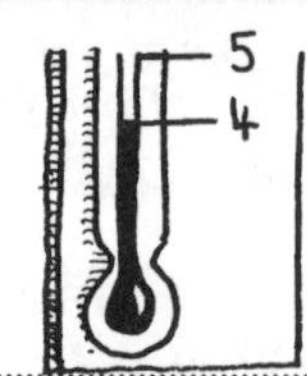

2 There are 130 seats in the upper part of a theatre. There are double this number in the stalls, the lower part.

How many seats are there in the stalls? **260**

3 Two buses take children on a school trip.

One bus has 68 children, the other has 27 children.

What is the total number of children on the trip? **95**

4 Jim says that the total of any two odd numbers and an even number is even.

Write four different examples of what Jim is saying.

Explain why Jim's claim is true.

5 A cinema has 480 seats. One afternoon only half of the seats have someone in them.

How many people are in the cinema on that afternoon? **240**

6 Carol has 35 books. Jelena has 46 more books than Carol.

How many books does Jelena have? **81**

7 ***If you add 39 to me and then subtract 28 from the answer you get 25.***

What am I?

14

Money problems 4

÷ £s by 2, 4, 5 or 10; –2D numbers; multiples of 50 total 1000

1 Beth buys four T-shirts for £12. Each T-shirt costs the same.

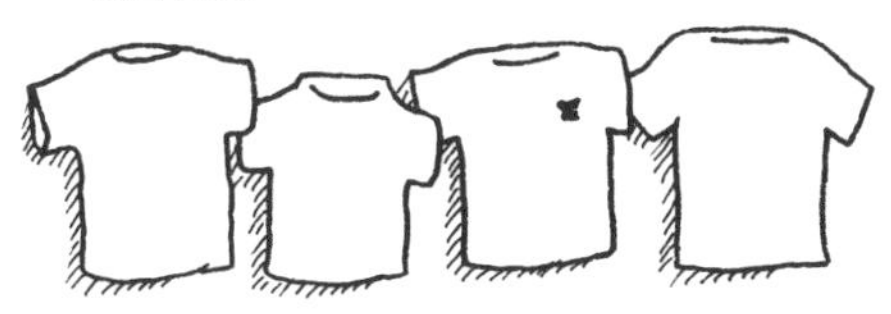

What is the cost of one T-shirt? **£3**

2 Sal wants to buy a £53 Game Girl. She only has £28.

How much more does she need? **£25**

3 Vinnie has £83 in two money boxes. The two boxes have the same amount in them.

How much is in each money box? **£41.50**

4 Mr Ross shares £37 equally between his five grandchildren.

How much does each grandchild get? **£7.40**

5 A holiday costs £550 per adult and £450 for each child. Trish and her mum go on the holiday.

How much does it cost Trish and her mum? **£1000**

6 Ten chicken sandwiches cost £27.

What does one chicken sandwich cost? **£2.70**

7

I am two numbers. The sum of my numbers is 1000. One of my numbers is 3 times my other number.

What am I?

250, 750

Number problems 11

×9 and 11; TU×U; +/– facts to 20

❶ Cans of juice are in boxes of 15.

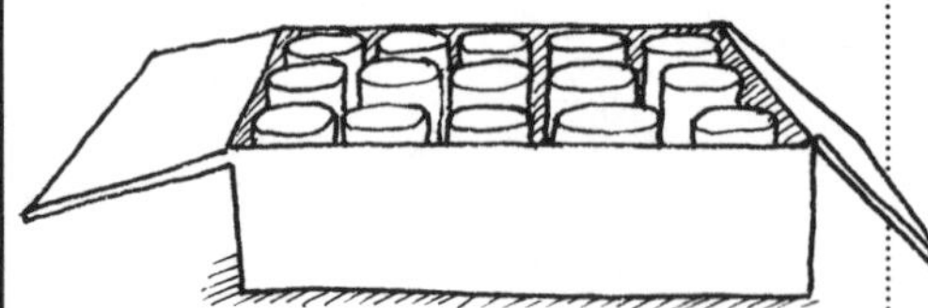

How many cans of juice are in 9 boxes? **135**

❷ When playing a rings game Roy scores 4 with his first ring and 12 with his second ring.

What is his total score with the two rings? **16**

❸ The seats in each row across a plane are in a two, three and two arrangement. The plane has 36 rows.

How many seats does the plane have? **252**

❹ There are 14 trees in a garden. Five of the trees have a disease and are cut down.

How many trees are left? **9**

❺ A box normally has 10 files. A special offer has 1 extra file in each box.

How many files are in 13 boxes with the special offer? **143**

❻ At a dinner 8 people are seated around each table. There are 24 tables at the dinner.

How many people are at the dinner? **192**

The sum of my two digits is 18.

What am I?

99

Number problems 12

fractions total 1; equivalent fractions; 2, 4, 10×, ÷5

1 Pansies are sold in packs of 8. Janice buys two packs.

How many pansies does Janice get altogether? **16**

2 Annette eats one-third of a cake.

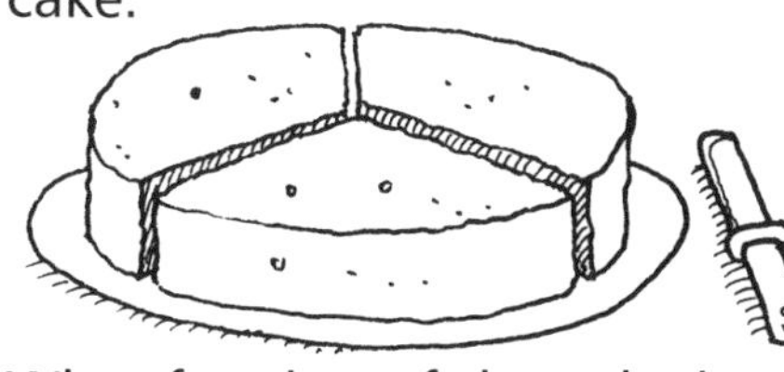

What fraction of the cake is left? $\frac{2}{3}$

3 Year 4 has 35 children. Their teacher puts them into teams of 5.

How many teams do Year 4 make? **7**

4 There are 20 candy sticks in each box. Anna eats one-fifth of them. Nigel eats two-tenths of them.

How many do Anna and Nigel each eat? **4**

5 Simon makes ladders. Each ladder has 10 rungs.

How many rungs does Simon need to make 17 ladders? **170**

6 Each display case has 7 model planes.

How many model planes are in four cases? **28**

7

I am a fraction. My top number is one less than my bottom number. Their sum is 3.

What am I? $\frac{1}{2}$

Number problems 13

multi-step problems

❶ A box of paints has 4 rows with 8 paints in each row. Claire buys a box and 10 spare paints.

How many paints did she buy? **42**

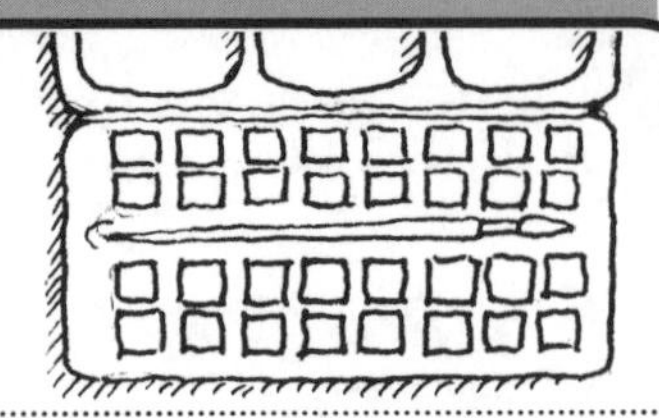

❷ Alfie buys a model car and a model plane to make. The car has two sets of 17 stickers and the plane has three sets of 15 stickers.

How many stickers are there altogether? **79**

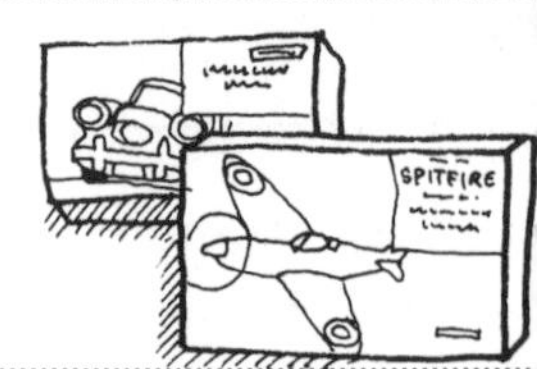

❸ Two books have 24 pages each and another book has 37 pages.

How many pages do the three books have altogether? **85**

❹ Rick shares his 50 sweets equally with his brother. He eats 12 of his sweets.

How many sweets does Rick have left? **13**

❺ There are 34 floors in an office block. The lift is at floor 21. It goes down 15 floors.

How many floors is it from the top? **28**

❻ A team is made up of 3 children and 2 adults.

How many people are there in 10 teams? **50**

❼

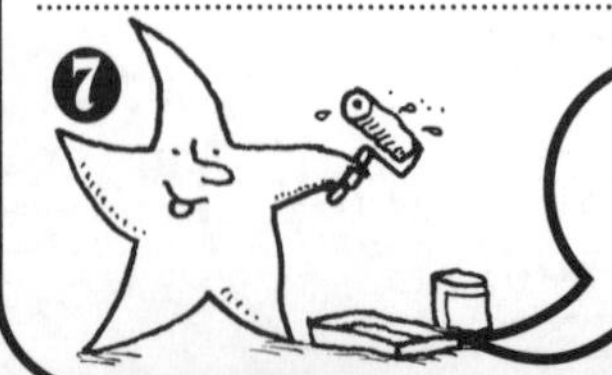

My tens digit is odd. The product of my two digits is 42.

What am I?

76

Review problems 4

review of previous content

1 At birth a blue whale calf weighs 3000 kg. Every day it puts on 60 kg of weight.

How much does it weigh after 10 days? **3600 kg**

2 A golden eagle chick is 13 cm long. When fully grown it is 88 cm long.

How many centimetres has a chick grown when it is an adult? **75 cm**

3 An olive tree lives for 3000 years. A bristle cone pine lives for 5000 years.

For how many more years does a bristle cone live than an olive tree? **2000 years**

4 A ferry can carry 3000 passengers. A jumbo jet can carry 500 passengers.

How many full jumbo jets are needed to fill a ferry? **6**

5 A Japanese Bullet train travels 200 miles in one hour. Stephenson's Rocket engine went 30 miles in one hour.

In one hour, how many more miles does the Bullet train go than did the Rocket? **170 miles**

6 A team of athletes flies to the Olympic Games. Of these, 329 fly on an airbus and 520 fly on a jumbo jet.

What is the total number of athletes on the two planes? **849**

7

I am a three-digit number. I am a multiple of 50. My first two digits are even and their difference is 6.

What am I?

600

Number problems 14

×100;
+/–10/100/1000

1 There are 294 children at Brimore School. Ten new children come to the school.

How many children are at the school now? **304**

2 Each Zarty tube has 58 chocolate drops. There are 100 tubes in each box.

How many chocolate drops are in a box? **5800**

3 Mr Wigton has 850 magazines to deliver to shops. At the first shop he leaves 100 magazines.

How many magazines has he left to deliver? **750**

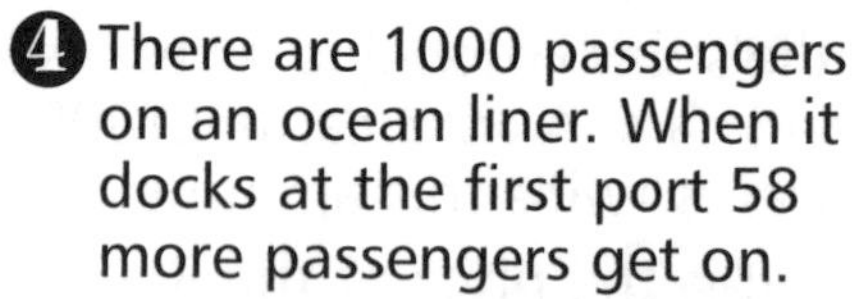

4 There are 1000 passengers on an ocean liner. When it docks at the first port 58 more passengers get on.

How many passengers are on the boat now? **1058**

5 A book has 340 pages of print. A printer makes 100 copies of the book.

How many pages does he print? **34 000**

6 A builder wants to build 41 houses. The council makes him decrease his plans by 10 houses.

How many houses is the builder allowed to build? **31**

7

If you subtract 6 hundreds from my number and then subtract 3 tens from the answer you end with 2.

What am I?

632

Money problems 5

+/–10/100/100; doubles

1 Imran collects £450 for a charity. Debbie and her friend collect twice as much.

How much do Debbie and her friend collect? **£900**

2 For two people sharing a room the cost of a holiday is £513 each. The cost of a single room is £100 extra.

What is the cost of a single room on the holiday? **£613**

3 Jodie's dad is looking at two second-hand cars. A Ford car is £1800 and a Rover is twice this amount.

What is the price of the Rover? **£3600**

4 A Flower Show takes place in a school hall. People pay £927 to enter the show.

The hall costs £100 to hire.

How much money does the show make? **£827**

5 A Scouts' and Guides' troop have £895 towards a target of £2000. A kind person gives them another £1000.

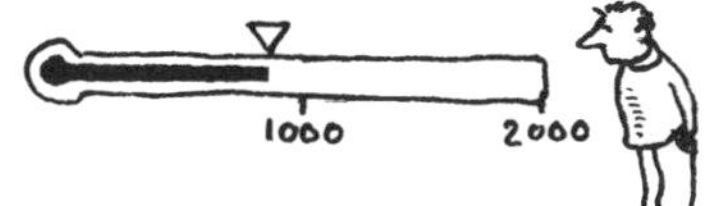

How many £s do they need to reach their target? **£105**

6 Alex's family take £200 on holiday. They spend £10 on ice cream and then go to the bank for another £100.

How much money have they now? **£290**

7

I am a multiple of 100.
I am more than double 250.
I am less than half of 1400.

What am I?

600

Number problems 15

facts to +/–; +/– neare[st] multiple of 10; + multiples of 10; halves

1 Model cars are packed in boxes of 100. A van driver unloads 400 boxes. He still has 900 boxes in his van.

How many boxes did he have to start with? **1300**

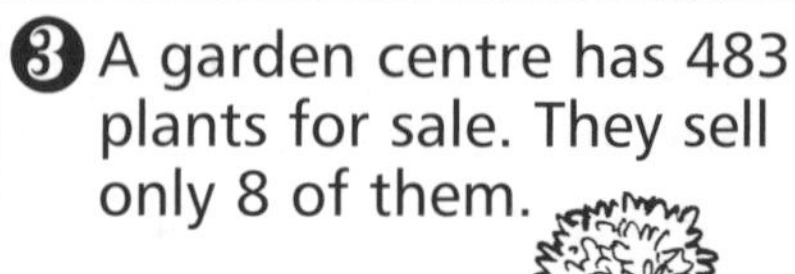

2 On Sunday 317 people watched a tennis match. This is 29 more than watched a match on the Saturday.

How many watched the match on the Saturday? **288**

3 A garden centre has 483 plants for sale. They sell only 8 of them.

How many plants are left for sale? **475**

4 There are 6200 tickets for sale for a cup final. Half of the tickets are sold before the game.

How many tickets are still for sale? **3100**

5 A grey squirrel collects 76 nuts. This is 39 fewer than a red squirrel collects.

How many nuts does the red squirrel collect? **115**

6 A train has three carriages. It has 30 people in the first carriage, 80 in the second and 70 in the third.

How many passengers does the train have altogether? **180**

7

I am a two-digit number.
I am 1 less than a multiple of 10.
The sum of my digits is 15.

What am I? **69**

Capacity problems 1

$\frac{1}{2}$, $\frac{1}{4}$, $\frac{1}{10}$ *of 1L in mL; pints; +/– facts to 20*

1 Jill drinks one-half of a litre bottle of milk.

How many millilitres are left in the bottle? **500 mL**

2 At the end of a day a garage has used 13 pints of oil. It has 2 pints left.

How many pints of oil did the garage have to start with? **15 pints**

3 A cask of wine holds 50 litres. Seven bottles are filled from a full cask. Each bottle has 700 mL.

How many millilitres are left in the cask? **45100 mL**

4 A litre bottle of orange is shared equally between four children.

How many millilitres of orange does each child get? **250 mL**

5 Each bottle has 1 pint of milk. A crate of 12 bottles is dropped, breaking 4 bottles.

How many pints of milk are left? **8 pints**

6 Ten identical glasses are filled to the same level using 1 litre of juice.

How many millilitres of juice is in each glass?
100 mL

7

I am a number of millilitres. I am more than $\frac{1}{10}$ litre. I am less than $\frac{1}{4}$ litre. The sum of my 3 digits is 19.

What am I?

190 mL

Measures problems 3

pairs total 100; multiples of 50 total 1000

1 A bunch of bananas weighs 1 kg. Two of the bananas, each weighing 250 g, are eaten.

What is the weight of the bananas that are left? **500 g**

2 Will, a record breaker, hops on his left leg for 35 minutes and on his right leg for 65 minutes.

How long does he hop for in hr and min? **1 hr 40 min**

3 Gaby runs 10 times round a track for a total of 2.5 km.

How many metres is once round the track? **250 m**

4 A bottle of vinegar holds 300 mL.

How much in L and mL do five bottles hold? **1 L 500 mL**

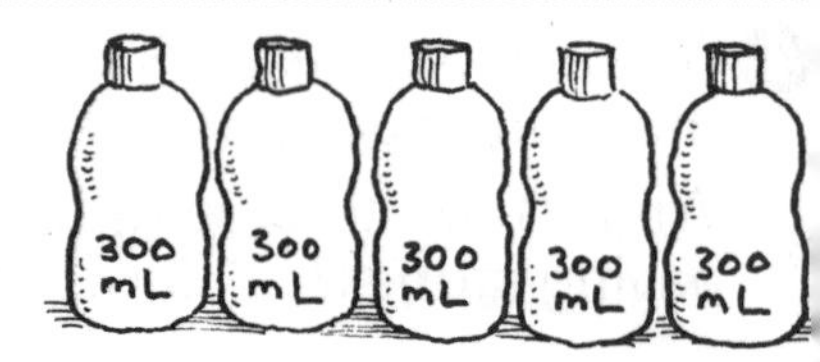

5 Joe practises standing jumps. He jumps 450 mm and then 550 mm.

How many metres has he jumped altogether? **1 m**

6 Moira catches a train at 11:30 a.m. The journey lasts 1 hr 15 min. She then gets another train and arrives at Stockley at 2:45 p.m.

How long was the second train journey? **2 hr**

7

I am the first number that is in both the count on in 3s and count on in 7s sequences.

What am I? **21**

Measures problems 4

angles; ×/÷ 10/100

1 A ship is travelling north. It changes direction anti-clockwise towards north-west.

Through what angle does it turn? **45°**

2 A case has 100 apples. The average weight of an apple is 150 grams.

How many kilograms do the apples in the case weigh? **15 kg**

3 A train on a turntable is facing south-west. The turntable turns 180°.

In what direction is the train now facing? **NE**

4 Paving stones are 75 cm square. Ten of them are placed in a straight line.

What is the total length of the 10 paving stones? **7.5 m**

5 There are 10 new cars in a show room. A total of 25 litres of petrol is put equally into the 10 cars.

How much petrol is put into each car? **2 L 500 mL**

6 A plane changes direction clockwise one right angle to fly south.

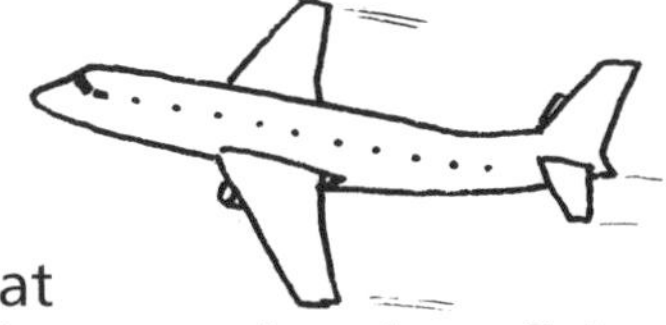

In what direction was the plane flying before it changed? **East**

7

I am an odd multiple of 5.
The sum of my two digits is 12.

What am I?

75

Review problems 5

1. There are 160 children at Dene Primary School. The children are divided equally into 4 teams for the sports.
   How many children are in each team? **40**

2. Half of the children in the Blue team are girls.
   How many boys are there in the Blue team? **20**

3. Each team is looked after by 5 teachers.
   How many teachers altogether look after the 4 teams? **20**

4. Each child is allowed only 2 people to come and watch.
   What is the maximum number of people coming to watch? **320**

5. Adults pay 10p and children 5p to watch. A hundred adults and 100 children come.
   How many £s do they pay in total? **£15**

6. The running track is 50 metres long.
   How many times must children run along the track to cover 100 metres? **2**

7. Which team has the most points on the score board? **Yellow**
   Which team has the least points on the score board? **Blue**

8. What is the order of the runners in the picture?
   4, 2, 1, 3 or G, B, R, Y

9. For each race 20 points are for 1st place, 10 for 2nd place and 5 for 3rd place. Red team has won 6 races and come 2nd in 5 races.
   In how many races has it come 3rd? **1**

10. There are 25 races altogether. Each runner finishing 1st, 2nd or 3rd gets a prize.
    How many prizes should there be altogether? **75**

11. How many races have already taken place?
    (Hint: In each race a total of 35 points is awarded). **21**

Number problems 16

+/– facts to 20; ×2, 3, 4 and 5

1. Every pack has 40 bird cards.
   How many cards are in 2 packs? **80**

2. There are 15 ducks feeding on a pond. Nine of them fly away.
   How many ducks are left feeding? **6**

3. Each bus has 70 children.
   How many children are there altogether on the five buses?
   350

4. The librarian puts 60 books on each of the four shelves.
   What is the total number of books he puts on the shelves? **240**

5. The 9 children in the playground are joined by 8 more children.
   How many children are then in the playground? **17**

6. Beth has 50 stickers. Amanda has three times as many stickers as Beth.
   How many stickers has Amanda?
   150

7. *I am the third number that is in both the count on in 3s and count on in 4s sequences.*

   What am I?

   36

Money problems 6

×2 and 4; ÷5 and 10; round up/down after ÷

1 Five packets of cheese and potato pie cost £4.95.

What is the price of one packet? **99p**

2 This is a pack of MacDill's Microchips.

What is the cost of four packs in £s and p? **£3.96p**

3 A 2 kg pack of chicken thighs costs £1.99. A 4 kg pack costs 10p less than two 2 kg packs.

What is the price of a 4 kg pack? **£3.88**

4 A shop has a special offer. The cost of shopping is rounded down to the nearest £.

Sinead spend £27.57.

How much does Sinead actually pay? **£27.00**

5 The price of a pack of 10 spring rolls is £1.99.

What is the cost of one spring roll to the nearest p? **20p**

6 Renee buys 4 packets of fish fingers and 2 grilled chickens. She has a £1.50 OFF voucher.

How much does she pay? **£16.44**

7 ***The difference between my two digits is 1. I am a multiple of 4 and more than 20. My tens digit is odd and 1 less than my units digit.***

What am I?

56

Number problems 17

×9 and 11; +2D numbers; ÷2 and 4; TU÷U; round up/down after ÷

1 A Landrover with a trailer takes 14 sheep to market. It does this 9 times in one week.

How many sheep go to market in the one week? **126**

2 On a farm there are 84 chickens. They lay 37 eggs. All the eggs hatch.

How many chickens are there now? **121**

3 David has to carry 23 boxes upstairs. He can only take two boxes each time.

How many times must David go up the stairs to take all 23 boxes? **12**

4 Each pack has 16 coloured crayons.

How many crayons are in 11 packs? **176**

5 A school takes 84 children on a trip to a farm. They are grouped in sixes.

How many groups are there altogether? **14**

6 Jim puts 38 cans into boxes. Each box can hold 4 cans.

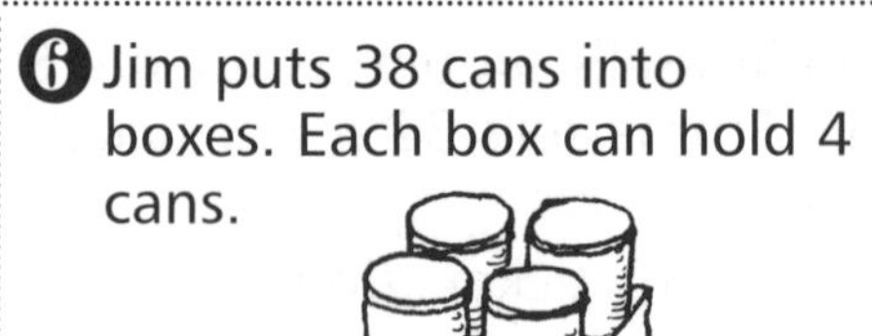

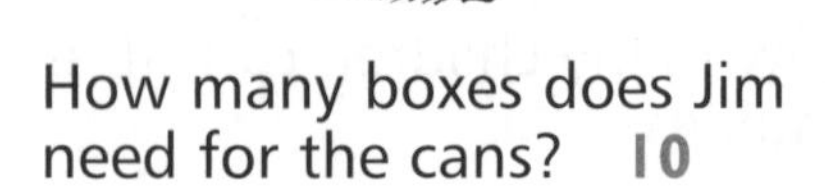

How many boxes does Jim need for the cans? **10**

7

My 4 digits are in order, smallest first. The 4 digits are in the count on in 2s sequence.

What am I?

2468

Number problems 18

proportion;
–2D numbers;
×6 and 8;
$\frac{1}{2}$ as decimal

1 In every pack of candy sticks there are 2 animal cards. Rick buys 5 packs.

How many animal cards does Rick get? **10**

2 Each book has six 1st class stamps.

Ged buys seven books.

How many stamps does he buy?
42

3 Rehana uses her calculator to find the answer to '1 divided by 2'.

What decimal will appear in the display of the calculator? **0.5**

4 A netball team plays one match every seven days.

How many matches will the team play in 35 days? **5**

5 Khadim sells 94 tickets for the Summer Fair. Emma sells 37 fewer tickets than Khadim.

How many tickets does Emma sell? **57**

6 At school dinner eight children sit around each table.

How many children can sit around 6 tables? **48**

7

I am between 61 and 99. I am a multiple of both 4 and 5.

What am I?

80

Time problems 3

timetables; calendars; ×3, 4 and 9

❶ A plane is due to leave at 9:30 a.m. and arrive at 10:40 a.m. It departs 5 minutes late, but arrives on time.

How long did it take? **1 hr 5 min**

❷ A train normally takes 25 min for a journey. Because of snow on the line it takes three times as long.

How long does the journey take? **1 hr 15 min**

❸ Chris' birthday is on 28th June. He goes on holiday on 26th July.

How many weeks is it from his birthday to his holiday? **4 weeks**

❹ Terri got to the bus stop at 1:05 p.m. She missed the bus by 10 minutes.

At what time did the bus leave the bus stop? **12:55**

❺ Kath is training for a race. She swims for 20 minutes and rests for 15 minutes. She does this nine times each day.

For how long does she swim each day? **3 hr**

❻ Alice has a sleep-over at her friend's house from 11:00 a.m. on Tuesday until 11:00 a.m. on Saturday.

How many hours is Alice at her friend's? **96 hr**

❼ 

I am between 45 and 65.
I am in both the count on in 2s and the count on in 7s sequences.

What am I? **56**

Number problems 19

+/–2D numbers; ×3 and 5; HTU+/–HTU

1 A village school has 96 boys and 60 girls.

How many children are there altogether? **156**

2 One afternoon 282 people watched a film. In the evening 347 saw the same film.

Altogether how many people saw the film? **629**

3 In Year 4 there are 75 children. Of these, 48 bring a packed lunch.

How many children do not have a packed lunch? **27**

4 On each bus there are 45 people.

How many people are on the three buses? **135**

5 A city has 244 ambulances. On Tuesday 186 of them were taking patients to hospital.

How many ambulances were not in use? **58**

6 A factory is building 53 rally cars. It needs five wheels for each car.

Altogether how many wheels does the factory need? **265**

7 

My 4 digits are all different. They are odd and in order, largest first. The sum of my first and last digits is 10. The sum of my middle two digits is also 10.

What am I?

9731

Review problems 6

review of previous content

1 Aisha gets 85 drinking glasses ready for a party. Only 68 of them are used.

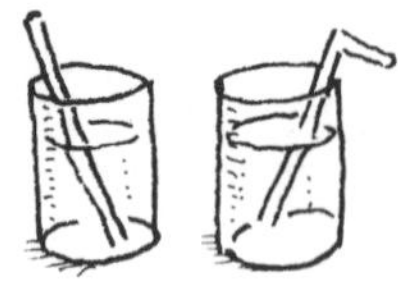

How many glasses are not used? **17**

2 A can of beans weighs 300 g. Two cans of beans are cooked and equal amounts put onto four plates.

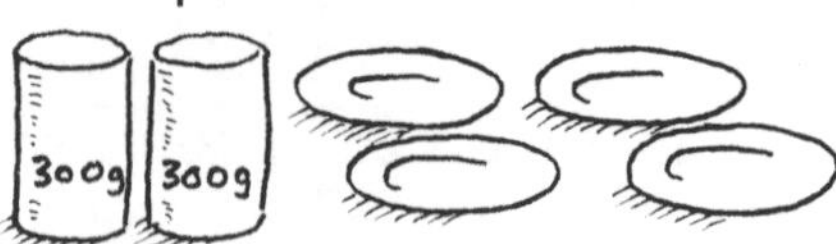

How many grams are on each plate? **150 g**

3 From London a train takes 2 hr 50 min to get to Leeds. It takes another 2 hr 15 min to get to Edinburgh.

How long does it take to get from London to Edinburgh? **5 hr 5 min**

4 Mrs Cook buys each of four children a colouring book and a pack of crayons.

How much do the books cost her? **£10.00**

5 Shola has gone 84 miles when he sees the signpost.

How far will he have travelled if he continues to Rill? **120 miles**

6 How many small washing up bottles hold the same as three large bottles? **4**

750 mL

1 Litre

7

I am two numbers. The sum of my numbers is 100. The difference between my numbers is 2.

What am I?

49, 51